Green Apple

J. M. Barrie

Peter Pan

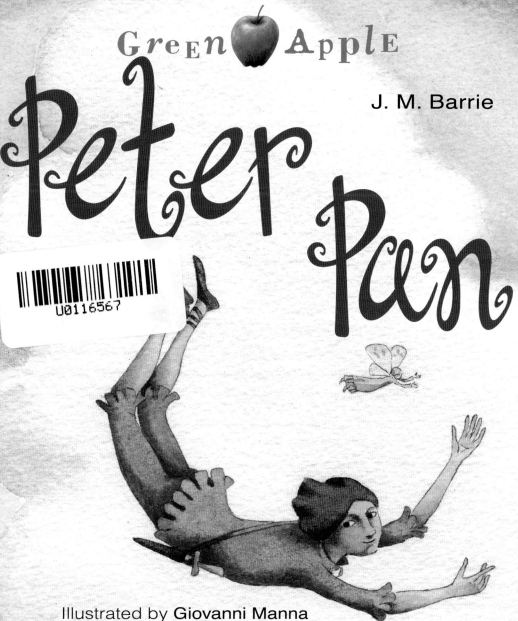

Illustrated by **Giovanni Manna**

Retold by **Gina D. B. Clemen**

Editors: Victoria Bradshaw, Eleanor Donaldson
Design and art direction: Nadia Maestri
Computer graphics: Simona Corniola
Picture research: Laura Lagomarsino

Picture credits:
© Bettmann/CORBIS: 4; Great Ormond Street Hospital for Children NHS Trust: 5;
Private Collection/Bridgeman Art Library: 31; © Blue Lantern Studio/CORBIS: 55;
MIRAMAX/Album: 61.

We would be happy to receive your comments and suggestions,
and give you any other information concerning our material.
http://publish.commercialpress.com.hk/blackcat/

CISQ

CISQCERT
**TEXTBOOKS AND
TEACHING MATERIALS**
The quality of the publisher's
design, production and sales processes has
been certified to the standard of
UNI EN ISO 9001

ISBN 978 962 07 0445 1

The CD contains an audio section (the recording of the text) and a CD-ROM section (additional fun
games and activities that practice the four skills).
- To listen to the recording, insert the CD into your CD player and it will play as normal. You can also
  listen to the recording on your computer, by opening your usual CD player program.
- If you put the CD directly into the CD-ROM drive, the software will open automatically.

SYSTEM REQUIREMENTS for CD-ROM	
PC:	**Macintosh:**
- Intel Pentium II processor or above (Intel Pentium III recommended)	- Power PC G3 processor or above (G4 recommended)
- Windows 98,ME,2000 or XP	- Mac OS 9.0 with CarbonLib or OSX
- 64 Mb RAM (32 Mb RAM Memory free for the application)	- 64 Mb RAM (32 Mb RAM free for the application)
- SVGA monitor 800x600 screen 16 bit	- 800x600 screen resolution with thousands of colours
- Windows compatible 12X CD-ROM drive (24X recommended)	- CD-ROM Drive 12X (24X recommended)
- Audio card with speakers or headphones	- Speakers or headphones
All the trademarks above are copyright.	

Contents

Special Features:

The text is recorded in full.

These symbols indicate the beginning and end of the extracts linked to the listening activities.

Let's meet Sir James M. Barrie

Name	James Matthew Barrie
Date and place of birth	9 May 1860 in Kirriemuir, Scotland
Plays and novels	*The Little Minster* (1891), *The Little White Bird* (1902), *Peter Pan* – the play (1904), *Peter Pan in Kensington Gardens* (1906), *Peter Pan and Wendy* – the novel (1911) and many others
Awards	title of 'Sir' in 1911
Date of death	19 June 1937

J. M. Barrie and Great Ormond Street Hospital for Children

J. M. Barrie likes children very much. He wants children to be well and happy.

In London there is a hospital for sick children called Great Ormond Street Hospital for Children. Barrie often visits this hospital because he wants to help the children.

Barrie is a very generous man. In 1929 he makes an important decision: all the money from book sales, films and theatre shows of *Peter Pan* goes to the hospital. This also continues after Barrie's death. This money helps the children, the hospital, the doctors and the nurses.

Peter Pan is performed once a year at the hospital. This is a very happy event for the children.

Great Ormond Street Hospital for Children in 1929.

The Characters

Captain Hook

The Lost Boys

Tinker Bell

Wendy

John

Michael

Peter Pan

Mr and Mrs Darling

BEFORE YOU READ

1 **Here are some words from the story. Do you know them? Match a word from the box to the correct picture.**

| alarm clock | arm | cheek | fairy | hook | kennel | mermaid |
| paw | sewing basket | shadow | shoulder | tail | tear |

1 ....

2 ....

3 ....

4 ....

5 ....

6 ....

7 ....

8 ....

9 ....

10 ....

11 ....

12 ....

13 ....

CHAPTER **ONE**

The Nursery [1]

endy, John and Michael Darling live in a lovely house in London. They have got a big, sunny nursery. There are colourful pictures and a big clock on the wall. There are toys here and there. The Darlings are a happy family. Mr Darling and Mrs Darling love their children very much. Wendy is the first child, John is the second and Michael is the third.

The children's nanny [2] is called Nana and she is a big Newfoundland dog! Her kennel is in the nursery and she is a wonderful nanny. She loves the children and the children love her.

One evening Mr and Mrs Darling want to go to a dinner party. They have their best clothes on.

---

1. **nursery** : a child's bedroom or playroom.
2. **nanny** : this person (here, a dog) looks after children.

'Nana, it's time to put the children to bed,' says Mrs Darling.

Nana goes to the bathroom. She turns on the hot water for Michael's bath. She puts her paw in the water to check the temperature. It's perfect!

'I don't want to have a bath!' says little Michael.

But Nana is a firm [1] nanny and Michael has his bath.

Then Nana gives the children their pyjamas. Now they are ready for bed.

Mrs Darling comes into the nursery and smiles. 'Good work, Nana! I see the children are ready for bed.' Nana wags [2] her big tail.

Suddenly there is a noise. Mrs Darling sees a young boy outside the nursery window. She is very surprised. Nana barks [3] and shuts the window quickly. The boy's shadow falls on the floor. The young boy flies away.

'Who's there?' asks Mrs Darling. She opens the window and looks outside, but she sees nothing. Then she sees the boy's shadow on the floor and says, 'Poor boy, this is his shadow. Let's put it in the drawer.' [4]

The children are in bed. Mr Darling takes Nana to the garden. Then he goes to the sitting room and waits for Mrs Darling.

Mrs Darling sings to the children and kisses them. She is a perfect mother. Soon the three children are sleeping. Mr and Mrs Darling go to their bedroom. They put on their coats and go to the dinner party.

---

1. **firm** : good at taking control.
2. **wags** : moves from side to side.
3. **barks** : makes a short, loud noise.   4. **drawer** : [drɔː]

UNDERSTANDING THE TEXT

KET

1 Are these sentences 'Right' (A) or 'Wrong' (B)? If there is not enough information to answer 'Right' (A) or 'Wrong' (B), choose 'Doesn't say' (C). There is an example at the beginning (0).

0 Wendy, John and Michael Darling live in London.
    (A) Right   B Wrong   C Doesn't say

1 Mr Darling works in a bank.
    A Right   B Wrong   C Doesn't say

2 The children's nanny is a big woman called Nana.
    A Right   B Wrong   C Doesn't say

3 Mrs Darling sees a young boy outside the nursery window.
    A Right   B Wrong   C Doesn't say

4 Mrs Darling puts the shadow in her bedroom.
    A Right   B Wrong   C Doesn't say

5 Mrs Darling sings to the children.
    A Right   B Wrong   C Doesn't say

6 Wendy takes Nana into the garden.
    A Right   B Wrong   C Doesn't say

7 The dinner party starts at 8 p.m.
    A Right   B Wrong   C Doesn't say

2 VOCABULARY – ROOMS IN THE HOUSE
Read the descriptions of some rooms in a house. Write the name of each room in the correct place.

hall	nursery	bathroom	cellar	bedroom
	kitchen	sitting room	attic	

1 You prepare meals here.   ....

2 Children play and sleep here.   ....

3 You have a bath here.   ....

4 You read and watch TV here.   ....

5 Your parents sleep here.   ....

3 In which rooms can you find these objects? Write the name of each room in the correct place.

| attic | dining room | sitting room | bedroom |
| hall | nursery | bathroom | kitchen |

1 milk, bread, cups ....
2 soap, water ....
3 toys, children's beds ....
4 TV, stereo, sofa ....
5 beds, clothes ....

T: GRADE 2

 4 **SPEAKING**
Topic – Family
Wendy is the first child. She has got two young brothers. Talk about your family. Use these questions to help you.

- Have you got any brothers or sisters? How old are they and what are their names? Describe what they look like.
- Are you an only child, the first child, the second child, etc.?
- Have you got any cousins? Answer the same questions for them.

 5 **SPEAKING AND WRITING – PETS**
A dog looks after the Darling children. Do you have any pets? What do you do to look after them? Do they sometimes look after you?
Write about your favourite animal (60-70 words) saying why you like this animal and what this animal does that is different or special. You can use the example to help you.

My favourite animal is a parrot. My grandma has got a parrot. His name is Rainbow because his feathers are many different colours, like a rainbow. I love parrots because they are beautiful birds and they are clever. Rainbow talks to me. He can say about 50 different words.

CHAPTER **TWO**

The Shadow

he children are sleeping and dreaming.  Suddenly the window opens. A small ball of light enters the nursery and flies around. It is a lovely fairy called Tinker Bell. She is looking for something. After a moment a young boy enters the nursery and says, 'Tink, where are you? Please find my shadow.'

Tinker Bell finds his shadow in the drawer and gives it to him. 'Now I can stick the shadow to my feet with some soap,' he thinks. He tries and tries again, but he can't. He is very confused and starts crying.

Wendy wakes up and sees the boy but she is not afraid. His clothes are made of leaves.

'Little boy, why are you crying?' Wendy asks.

The boy takes off his cap [1] and asks, 'What's your name?'

'Wendy Moira Angela Darling. What's yours?'

'Peter Pan.'

'Is that all?'

'Yes!' says Peter. Then he thinks, 'My name is very, very short.'

Wendy looks at his shadow and asks, 'Can I help you with your shadow?'

'Yes, please!' says Peter.

Wendy gets her sewing basket and sews on [2] Peter's shadow.

After a few minutes she says, 'Finished! Now you have your shadow again.'

1. **cap** : type of hat.
2. **sews on** : attaches with a needle  and cotton.

Peter looks at the floor and sees his shadow. He is very happy and dances around the room.

'Oh, Wendy, you're wonderful!' says Peter.

'Do you really think so?' asks Wendy.

'Yes,' says Peter.

Wendy smiles and gives Peter a kiss on the cheek.

'Oh!' says Peter. 'How nice!'

'How old are you, Peter?' asks Wendy.

'I don't know, but I'm young. I don't want to grow up. I always want to be a boy and have fun.'

Peter looks around the room for his fairy. He hears a noise and looks in a drawer. Tinker Bell flies out. Wendy is delighted [1] to see a fairy, but Tinker Bell is afraid. She hides behind the big clock.

'Where do you live, Peter?' asks Wendy.

'I live in Neverland with the Lost Boys,' [2] says Peter.

'Neverland? The Lost Boys? Who are they?' asks Wendy.

'The Lost Boys haven't got a mother or father. They are alone in the world and they live in Neverland. I am their Captain. In Neverland we

1. **delighted** : very happy.
2. **Lost Boys** : boys without a home.

fight the pirates. We also swim in the lagoon [1] with the beautiful mermaids. Fairies live in the trees in the forest. The fairies are my friends,' says Peter.

'Oh, what fun!' says Wendy.

'I must go back now. I must tell the Lost Boys a story. They love stories,' says Peter.

'Don't go away! I know a lot of stories,' says Wendy.

'Then come with me, Wendy. You can tell us stories. We all want a mother. Please come,' says Peter.

'But I can't fly,' says Wendy.

'I can teach you to fly,' says Peter.

'Can you teach John and Michael to fly, too?'

'Yes, of course,' says Peter.

'John! Michael! Wake up! This is Peter Pan. He's from Neverland. It's a beautiful place,' says Wendy.

John and Michael are very surprised.

'We can go there with him. But first we must learn to fly,' Wendy says.

Wendy, John and Michael are very excited. They try to fly but fall on the beds and on the floor.

'No, no,' says Peter. 'Here is some fairy dust.' [2] He puts some fairy dust on their shoulders.

'Now try again,' says Peter.

'Look, I can fly!' says Wendy.

'I can too,' says John.

'Me too,' says little Michael.

'Tink, show us the way to Neverland,' says Peter.

---

1. **lagoon** : an area of salt water separated from the sea.

2. **fairy dust** : dust is a powder. Fairy dust is a type of magic powder.

They follow Tinker Bell and fly out of the nursery window. In the garden Nana looks at the sky and barks.

Mr and Mrs Darling return from the dinner party. They go into the nursery, but it is empty!

UNDERSTANDING THE TEXT

1 **Choose the correct answer (A-I) for each question.**

1 ☐ Who enters the nursery when the children are sleeping?
2 ☐ Where is Peter's shadow?
3 ☐ Where does Peter live?
4 ☐ What does Wendy do to help Peter?
5 ☐ Why is Peter happy?
6 ☐ Who is happy to see a fairy?
7 ☐ Why must Peter return to Neverland?
8 ☐ Who must learn to fly?
9 ☐ Where is Nana?

A Wendy, John and Michael.
C In Neverland.
E Wendy.
G She is in the garden.
I Because he must tell the Lost Boys a story.

B Because he has his shadow again.
D Tinker Bell and Peter Pan.
F In the drawer.
H She sews on his shadow.

2 **CHARACTERS**
These shadows belong to characters in the story. Who do they belong to?

1 .... 2 .... 3 .... 4 ....

3 **VOCABULARY**
Imagine a fairy! Circle the adjectives that best describe one. Use them to write a sentence about Tinker Bell.

kind	small	happy	old	clever	good
fat	magic	beautiful	funny	bad	young

4 LISTENING

Listen to the following conversation. Choose the correct answer, A, B or C.

1　The nursery is

A ☐ sunny.
B ☐ empty.
C ☐ full.

2　Perhaps the children are

A ☐ playing.
B ☐ sleeping.
C ☐ hiding.

3　Mr and Mrs Darling look

A ☐ under the beds.
B ☐ under the chairs.
C ☐ in the kennel.

4　The nursery window is

A ☐ locked.
B ☐ closed.
C ☐ open.

5　What time is it?

A ☐ midnight
B ☐ noon
C ☐ half past twelve

6　Mr Darling wants to

A ☐ call Nana.
B ☐ look in the garden.
C ☐ go to bed.

5 PREPOSITIONS

Complete the sentences with the correct preposition.

| behind | into | on | out | with |

1　The children fly .... of the nursery window.
2　Wendy gives Peter a kiss .... his cheek.
3　Tinker Bell hides .... a big clock.
4　'Let's go to Neverland .... Peter!' says Wendy.
5　Mr and Mrs Darling go .... the empty nursery.

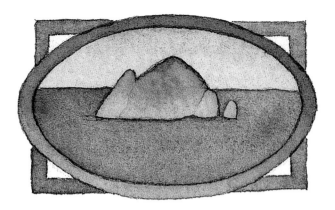

CHAPTER **THREE**

Neverland

endy, John and Michael fly over cities, towns, mountains, forests and seas.

Finally they see an island in the sea below them.

'Look, that's Neverland,' says Peter.

'Neverland!' say the children.

In Neverland the Lost Boys live in the forest in a secret underground home (*see the map of Neverland*). There are six Lost Boys: Slightly Soiled, [1] Tootles, Nibs, Curly and the Twins. They are waiting for Peter.

Suddenly they hear the voices of the pirates. Nibs is very brave. [2] He goes out, hides behind a tree and looks around him. He sees the horrible pirates. They are walking in the forest. They are big and ugly.

(*Their pirate ship is the* Jolly Roger.)

---

1. **Slightly Soiled** : this name means 'a little dirty'.
2. **brave** : courageous, not afraid of anything.

The pirates' captain is James Hook. He is a cruel pirate and a very bad man. He's got black eyes, black hair and a black beard. [1] He hates Peter Pan. He's only got one hand: the other is a hook! In the past Peter Pan cut [2] off Captain Hook's right arm during a fight. A crocodile ate [3] the arm. Now the crocodile follows Captain Hook everywhere because he wants to eat him. The crocodile has an alarm clock in its stomach! Captain Hook can always hear it.

'I know the Lost Boys live in this forest. We must find them and Peter Pan!' says Captain Hook.

'Tick, tock, tick, tock!' Captain Hook hears the alarm clock.

'Oh, no, the crocodile's coming to eat me!' says Captain Hook. He runs away and the pirates follow him.

Soon some Indians arrive in the forest. They are looking for the pirates. The pirates are their enemy. Tiger Lily is their leader. She is the beautiful daughter of the Indian chief. She loves Peter Pan. Tinker Bell and Wendy love him too. The Indians go away and the Lost Boys return to play in the forest.

Then Nibs looks at the sky and says, 'Look, there's a lovely white bird in the sky.'

'Is it really a bird?' the Lost Boys ask.

Tinker Bell says, 'Yes, yes, it's a bird. It's a Wendy bird. You must shoot it!' Sometimes Tink is a bad fairy. She knows it is Wendy, but she doesn't like her.

1. **beard** :
2. **cut** : the Past Simple of the verb 'to cut'.
3. **ate** : the Past Simple of the verb 'to eat'.

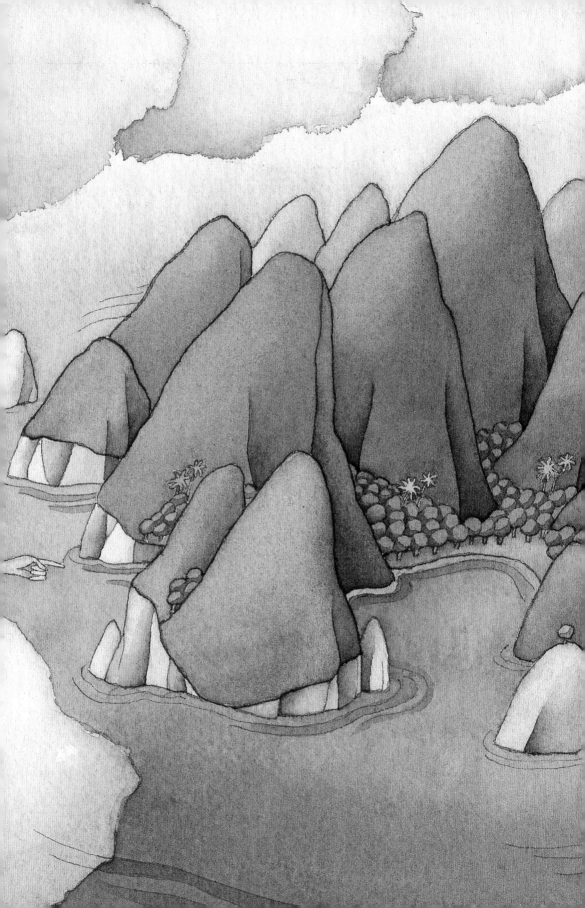

Nibs takes his bow and arrow and shoots
Wendy. Poor Wendy falls to the ground.
The Lost Boys see Wendy and say,
'She's not a bird! She's a lovely girl.'
Peter flies down with John and
Michael and asks, 'Where's Wendy?'
Tootles says, 'Here she is.'
Peter goes over to her and asks,
'Wendy, are you alright?'
Wendy slowly opens her eyes and
smiles.
'Yes, but I'm very tired,' she says.
The Lost Boys are sorry. They decide
to build her a little house.

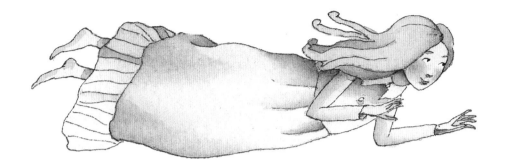

When the house is ready Wendy says, 'What a lovely little house! Thank you.'

'Can you be our mother now? Can you tell us bedtime stories before we go to bed?' asks Nibs.

'Of course,' says Wendy. 'Come in and I can tell you the story of Cinderella.'

They enter, sit down and listen to Wendy's story. It is a wonderful story.

Peter Pan is outside the house with his sword. [1] He wants to protect Wendy and the Lost Boys.

1. **sword** :

UNDERSTANDING THE TEXT

1 Are these sentences 'Right'? (A) or 'Wrong' (B)? If there is not enough information to answer 'Right' (A) or 'Wrong' (B), choose 'Doesn't say' (C). There is an example at the beginning (0).

0 Wendy, John and Michael fly all night.
A Right   B Wrong   Ⓒ Doesn't say

1 The Lost Boys live in a secret underground home in Neverland.
A Right   B Wrong   C Doesn't say

2 The pirates' captain is James Hook. He is a good man.
A Right   B Wrong   C Doesn't say

3 The crocodile wants to eat Captain Hook.
A Right   B Wrong   C Doesn't say

4 The crocodile is very old.
A Right   B Wrong   C Doesn't say

5 Nibs sees Peter, Tinker Bell and the children in the sky.
A Right   B Wrong   C Doesn't say

6 The Lost Boys build an underground house for Wendy.
A Right   B Wrong   C Doesn't say

7 Wendy tells the Lost Boys a story.
A Right   B Wrong   C Doesn't say

2 **WHAT TIME IS IT?**
The crocodile has an alarm clock in his stomach. Captain Hook can always hear it, but he can't tell the time! Look at the alarm clocks below and say what time they show. At what time do you usually set your alarm clock for?

1 ....   2 ....   3 ....

4 ....   5 ....   6 ....

LISTENING

Listen to part of the story again and fill in the missing words.

Soon some Indians .... in the forest. They are looking for the
.... The pirates are their .... Tiger Lily is ....
leader. She is the .... daughter of the Indian chief. She ....
Peter Pan. Tinker Bell and Wendy love him .... The Indians go
.... and the Lost Boys return to .... in the forest.

CAN YOU READ A MAP?

Look at the map of Neverland at the back of the book and answer the
questions. Tick (✓) the correct answer.

1 You are on the *Jolly Roger*. You want to go
   to the Mysterious River. You must go
   - A ☐ south.
   - B ☐ east.
   - C ☐ north.

2 Tiger Lily is in the Indian camp. She wants to go
   to Mermaids' Lagoon. She must go
   - A ☐ east.
   - B ☐ west.
   - C ☐ south.

3 John is in the Underground Home. He wants to
   go to Kidd's Creek. He must go
   - A ☐ south.
   - B ☐ west.
   - C ☐ east.

4 Michael and Nibs are at Slightly Gulch. They
   want to go to Kidd's Creek. They must go
   - A ☐ east.
   - B ☐ south.
   - C ☐ north.

5 Tinker Bell is at Kidd's Creek. She wants to
   go to the *Jolly Roger*. She must go
   - A ☐ south.
   - B ☐ east.
   - C ☐ west.

6 Peter and Wendy are on Marooner's Rock.
   They want to go to the Indian Camp.
   - A ☐ east.
   - B ☐ north.
   - C ☐ west.

Draw a map of your own secret island. Put as many places on it as
possible and give them a name. Are there any mountains? Is there a
river, or a lagoon? Is there a desert? Are there any underground
caves? When you finish, describe the island to another student.

Pirates

John wants to know more about pirates.
He asks Captain Hook these questions:

John: Who are pirates?

Hook: A lot of pirates are men but some pirates are women. Some pirates are cruel.

John: What do they do?

Hook: Pirates attack ships at sea. They take the treasure of the ship. Sometimes they kill the crew!

John: Oh! Where do they live?

Hook: They live on ships and on small islands.

John: Are there other names for pirates?

Hook: Yes, of course! Buccaneers, corsairs, rovers and filibusters.

John: Have pirate ships got flags?

Hook: Yes! A pirate's flag is usually black with a white skull on it. It's called 'The Jolly Roger'.

John: Tell me about a pirate's life on a ship.

Hook: It's very difficult. The food is bad and there isn't much water to drink. A pirate must always obey [1] his captain.

John: What does a pirate eat?

Hook: He eats dry biscuits, eggs, turtles, [2] fish and salted meat.

John: Are some pirates famous?

Hook: Oh, yes! Some pirates are *very* famous. They include Henry Morgan, Captain Kidd, Blackbeard, the Barbarossa Brothers and Black Bart!

John: Do pirates have a treasure chest? [3]

---

1. **obey** : do everything the captain wants.

2. **turtles** :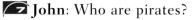

3. **treasure chest** :

Hook: Yes, they do. There are jewels, diamonds, gold, money and other precious things in the treasure chest.

John: What does a pirate wear?

Hook: A pirate captain wears beautiful clothes and a big hat. His clothes are colourful. Some captains wear jewels. The other pirates wear simple clothes: dark jackets, cotton shirts and comfortable trousers. Most pirates don't wear shoes!

John: How do pirates fight?

Hook: They fight with a sword and a pistol.

John: Oh! A pirate's life is exciting!

The Capture of the Pirate Blackbeard (1718) by Jean L. J. Ferris.

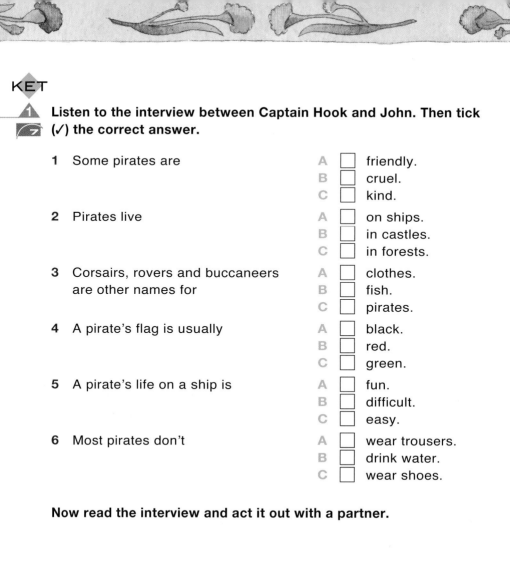

1 Listen to the interview between Captain Hook and John. Then tick
(✓) the correct answer.

1 Some pirates are

A ☐ friendly.
B ☐ cruel.
C ☐ kind.

2 Pirates live

A ☐ on ships.
B ☐ in castles.
C ☐ in forests.

3 Corsairs, rovers and buccaneers
are other names for

A ☐ clothes.
B ☐ fish.
C ☐ pirates.

4 A pirate's flag is usually

A ☐ black.
B ☐ red.
C ☐ green.

5 A pirate's life on a ship is

A ☐ fun.
B ☐ difficult.
C ☐ easy.

6 Most pirates don't

A ☐ wear trousers.
B ☐ drink water.
C ☐ wear shoes.

Now read the interview and act it out with a partner.

PROJECT **ON THE WEB**

In groups find out about pirates and their lives. Connect to the Internet
and go to www.blackcat-cideb.com or www.cideb.it . Insert the title or
part of the title of the book into our search engine. Open the page for
Peter Pan. Click on the internet project link. Go down the page until you
find the title of this book and click on the relevant links for this project.

Make a poster about a pirate (it can be a famous pirate or you can
invent one yourself!). Use the questions below to help you. You can
add pictures of the pirate, the ship or the flag, if you want.

1 What is his name?
2 What is his nationality?
3 What does he look like?

4 What's the name of his ship?
5 Describe his life as a pirate.

CHAPTER **FOUR**

The Mermaids' Lagoon

ne summer evening Peter, Wendy, John, Michael and the Lost Boys go to the Mermaids' Lagoon. Beautiful mermaids live here and they are Peter's friends. They swim and play in the blue lagoon. Then they sit on Marooner's Rock to comb [1] their long hair. They sit in the sun and laugh.

The children like the mermaids and John says, 'I want to catch one!' He tries, but the mermaid jumps into the water.

Peter says, 'It's very difficult to catch a mermaid.'

Suddenly someone says, 'Look, the pirates are coming!'

---

1. **comb** : [kaʊm] arrange or tidy hair with a .

A small boat with two pirates is coming to the lagoon. John, Michael and the Lost Boys jump off the rock and swim away. But Wendy stays with Peter. They hide behind the rock. Peter sees Tiger Lily. She is sitting in the small boat. Poor Tiger Lily is a prisoner [1] of the pirates.

'Let's leave her on this rock. When the sea rises, [2] she'll die!' says Smee. The two pirates laugh. It is already night and it is very dark.

Peter wants to save Tiger Lily and thinks of something intelligent. He imitates [3] Captain Hook's voice and says, 'Cut the ropes [4] and let her go! Do as I say, you idiots! Let her go!' The two pirates are amazed. [5]

'Can you hear Hook's voice?' asks Smee.

'Yes, but what do we do?' asks Starkey.

'We must obey him and cut the ropes,' says Smee.

They cut the ropes and Tiger Lily is free. She quickly jumps into the water and swims away.

1. **prisoner** : a prisoner is not free.
2. **rises** : becomes higher in level.
3. **imitates** : copies someone's actions or voice.
4. **ropes** :
5. **amazed** : very surprised.

Captain Hook sees everything and he is furious.

'That horrid Peter Pan! This time I must attack him,' he says.

He swims to the rock and fights with Peter. It is a long fight. The Captain hurts Peter with his hook, but Peter fights courageously. At last, Peter wins the fight and Hook swims back to the *Jolly Roger*.

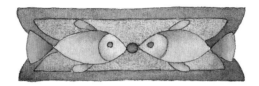

Peter is now alone on the rock with Wendy.

'The sea is rising and we are in great danger here. We must leave this rock,' says Peter.

'Oh, Peter, I am very tired and I can't swim or fly.'

He sees a big kite [1] with a long tail. It is flying slowly over the lagoon. He takes the tail of the kite and says, 'Wendy, hold on to this tail and fly away with the kite.' Wendy flies away.

'The sea is rising. I must fly away,' Peter thinks.

When he gets home everyone is happy to see him, especially Wendy.

1. **kite** :

UNDERSTANDING THE TEXT

KET

1 **Choose the correct words to complete the sentences. Tick (✓) the correct answer, A, B or C.**

1 The beautiful mermaids sit on Marooner's Rock and

- A ☐ sleep.
- B ☐ comb their hair.
- C ☐ sing.

2 John tries to catch a

- A ☐ mermaid.
- B ☐ fish.
- C ☐ pirate.

3 Tiger Lily is a prisoner of the

- A ☐ Indians.
- B ☐ Lost Boys.
- C ☐ pirates.

4 Peter wants to save Tiger Lily and imitates Captain Hook's

- A ☐ face.
- B ☐ voice.
- C ☐ laugh.

5 When Tiger Lily is free, Captain Hook is

- A ☐ surprised.
- B ☐ happy.
- C ☐ furious.

6 Peter and Wendy must leave the rock because the sea is

- A ☐ rising.
- B ☐ cold.
- C ☐ dirty.

7 Wendy flies away with

- A ☐ Peter.
- B ☐ the Never bird.
- C ☐ a kite.

2 **Which of the two pictures in Chapter Four do these sentences describe?**

1 'We must obey him and cut the ropes.'
2 John says, 'I want to catch one!'
3 Poor Tiger Lily is a prisoner of the pirates.
4 They swim and play in the Blue Lagoon.

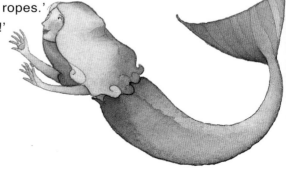

Complete this crossword puzzle.

ACROSS	DOWN

ACROSS

3 very angry

4

6

7 opposite of 'over'

8

10 opposite of quickly

11

DOWN

1

2

4 opposite of 'always'
5 colour of the lagoon
8 this person is not free

9

CHAPTER **FIVE**

The Underground Home

he Underground Home is a secret place. No one knows where it is. It is a happy, warm place. There is only one room with a big fireplace. [1] Tinker Bell has her tiny [2] room too.

Wendy is a perfect mother. She cooks and sews for everyone. She also tells beautiful bedtime stories. The Lost Boys are happy because they finally have a mother. John and Michael are happy because there is a new adventure every day.

Peter Pan is a perfect father. He brings home food and protects the family.

1. **fireplace** :

2. **tiny** : very small.

Wendy and Peter play with the children and laugh with them.

But one night something happens. Wendy tells the children this bedtime story:

'In the big city of London there are two parents. They're very sad because they can't find their three children. Every night they leave the nursery window open. They wait and wait for their children to return. But they don't return. Poor parents! They're very sad without their children.'

'Oh, Wendy, this is the story of our parents,' says John.

'Yes, it is,' says Michael.

Peter listens and says, 'Sometimes parents forget their children and other children take their place.'

Wendy is very surprised. 'Oh, no!' she says! 'Perhaps there are other children in our beds! John, Michael, we must go home!'

'Do we *really* have to?' ask John and Michael.

'Yes, we've *got to* [1] return home.'

The Lost Boys are sad and say, 'Oh, Wendy, please don't leave us!'

'Don't be sad. You can come and live with us in London,' says Wendy.

'Oh, how wonderful!' the Lost Boys say. 'We can have a real family.' They jump up and down with joy and they dance around the room.

But Peter is not happy. He is very serious and says, 'I'm not coming with you to London. I don't want to grow up. I want to be a boy forever.'

Everyone says goodbye to Peter. Outside, the pirates are waiting for them! The children come out of the underground

---

1. **we've got to** : we must.

home and the pirates capture [1] them. Then they take them to the *Jolly Roger*.

They don't make any noise. Peter doesn't know where they are.

He is sad without Wendy, John, Michael and the Lost Boys. He sits and thinks.

'Tap, tap, tap!' There is someone at the door.

'Who is it?' asks Peter.

He can hear the sound of little bells and opens the door. Tinker Bell flies in and says, 'The pirates have got Wendy, John, Michael and the Lost Boys! They are in danger. Let's help them!'

'I must save them. Come Tink, let's go to the *Jolly Roger*! This time I *must* attack Hook!'

1. **capture** : take them away as prisoners.

UNDERSTANDING THE TEXT

1 **Read the paragraph and choose the best word (A, B or C) for each space. There is an example at the beginning (0).**

The Underground Home (0) ...B......... a secret place. There is one room (1) .... a fireplace. Wendy cooks and sews for (2) .... She tells (3) .... bedtime stories. John and Michael are happy because there is (4) .... adventure every day. Wendy, John and Michael decide to (5) .... home. The Lost Boys want to go (6) .... London with Wendy, John and Michael.

Peter Pan doesn't want to grow (7) .... He (8) .... to stay in Neverland. Suddenly everyone hears a lot of noise. The Indians and the pirates (9) .... fighting. The fighting stops and the children go (10) .... The wicked pirates capture (11) .... Peter wants to (12) .... them.

0	A be	B (is)	C are
1	A with	B on	C in
2	A Peter	B someone	C everyone
3	A beautiful	B beautifully	C beauty
4	A any	B an	C a
5	A return	B remember	C forget
6	A at	B in	C to
7	A together	B up	C big
8	A wants	B want	C must
9	A is	B are	C am
10	A in	B outside	C inside
11	A us	B you	C them
12	A save	B see	C finish

2 **CHARACTERS**
Who does what? Find the answers in Chapter Five.

Who...

1 tells beautiful bedtime stories?   ....

2 brings home food?   ....

3 plays with the children?   .... and ....

4 must return to London?   .... , ...
                                    and ....

5 says goodbye to Peter? ....

6 captures the children? ....

7 is sad? ....

8 flies in? ....

KET

3. **What does Peter say to Wendy? Complete the conversation. For questions 1-6, write the correct letter A-G.**

0 *Wendy:* Good evening, Peter. *Peter:* .E....

1 *Wendy:* What time is it? *Peter:* ....

2 *Wendy:* Are you hungry? *Peter:* ....

3 *Wendy:* Let's have dinner! *Peter:* ....

4 *Wendy:* Where are the children? *Peter:* ....

5 *Wendy:* They must come home and wash their hands. *Peter:* ....

6 *Wendy:* Thank you, Peter. *Peter:* ....

A What a good idea! B That's OK.

C It's half past six. D I can call them.

E Hello, Wendy. F They're playing outside.

G Yes, I am.

4. **GROWING UP**

Look at this list. Which things do adults normally do? Are there any things that children do that adults can't?

pay the bills   cook the dinner   go to work

have their own money to pay for things   go to parties if they want to

stay up late   play computer games   play with toys   go to school

go roller skating   text friends   shout at their parents

5. **Peter doesn't want to grow up and become an adult. Would you like to be a child all your life?**

1 ☐ Yes, because then I don't have to go to work and earn money.

2 ☐ No, because it's more fun being an adult.

CHAPTER **SIX**

The Jolly Roger

here is a yellow moon in the night sky. The Jolly Roger is in the bay near Kidd's Creek. The children are on the pirate ship. They are prisoners of Captain Hook and his cruel pirates.

Captain Hook looks at them and says, 'This time it's Peter Pan or me! You idiots! Peter Pan can't save you now.' Hook laughs and then calls Smee. 'Smee, get the plank [1] ready!'

'Yes sir!' says Smee.

'Now listen to me,' says Hook. 'You must all walk the plank!'

'Walk the plank?' asks John.

'Yes! First you walk the plank and then you fall into the sea with the crocodile. It will eat you! Ha, ha!' laughs Hook. 'But I can save two of you. I want two young pirates. Who wants to be a pirate?'

The Lost Boys look at John. John looks at Michael and says, 'The life of a pirate is exciting. I don't want to walk the plank. I

---

1. **plank** : long, rectangular piece of wood. Pirates made people walk on this into the sea.

don't want to be food for the crocodile. Let's be pirates!' Michael looks at his brother. Then they look at Wendy. She doesn't like their idea.

Captain Hook laughs and moves his hook in front of their faces.

'Do you want to be pirates, yes or no?' he asks.

John and Michael say, 'Never!'

Captain Hook is angry and says, 'Then you must walk the plank and die!'

Wendy is afraid. She loves her brothers and the Lost Boys. She has tears in her eyes.

The boys stand near the plank and Wendy watches them. A pirate asks, 'Who's the first to walk the plank?'

At that moment there is a loud noise. 'Tick! Tock! Tick! Tock!'

Captain Hook's face is white. He says, 'The crocodile is here. He wants ME!' He runs to his cabin and hides there.

'Who is the first to walk the plank?' asks a pirate. 'Come on! Let's go! The crocodile is hungry.'

Suddenly Peter Pan appears on the pirate ship. Tinker Bell follows him. Wendy and the boys cheer. [1] They are very happy to see their young hero.

Hook and his pirates are furious. Hook takes his sword and says, 'I want to fight you, Pan! Tonight you will die!'

Hook fights with his long sword and with his hook. Peter fights courageously. He pushes Hook to the back of the ship. It is a terrible fight. John, Michael and the Lost Boys fight the pirates. After a long fight they throw the pirates into the sea.

Peter and Hook move all around the big ship. Their swords make a loud noise. Suddenly Peter takes Hook's sword and pushes him into the sea! Hook shouts, 'OH, NO!' He falls into the sea and into the mouth of the hungry crocodile.

'Oh, Peter, we are proud [2] of you!' says Wendy. She kisses him on the cheek. The boys cheer. Peter smiles and says, 'The *Jolly Roger* is ours now. Let's go home!'

---

1. **cheer** : shout happily.   2. **proud** : pleased and satisfied.

UNDERSTANDING THE TEXT

1 **Are these sentences 'Right' (A) or 'Wrong' (B)? If there is not enough information to answer 'Right' (A) or 'Wrong' (B), choose 'Doesn't say' (C). There is an example at the beginning (0).**

0 The children are prisoners of the Indians.
A Right   (B) Wrong   C Doesn't say

1 They must walk the plank.
A Right   B Wrong   C Doesn't say

2 The Lost Boys start crying.
A Right   B Wrong   C Doesn't say

3 Captain Hook is not afraid of the crocodile.
A Right   B Wrong   C Doesn't say

4 Peter Pan appears on the pirate ship at half past eleven.
A Right   B Wrong   C Doesn't say

5 The children are very happy to see Peter.
A Right   B Wrong   C Doesn't say

6 Peter and Captain Hook fight with bows and arrows.
A Right   B Wrong   C Doesn't say

7 Captain Hook falls into the sea and into the mouth of the crocodile.
A Right   B Wrong   C Doesn't say

8 The long fight finishes in the morning.
A Right   B Wrong   C Doesn't say

2 **LISTENING**

Listen to Wendy's letter and complete the spaces. Then complete the sentences. Match the numbers with the letters to give the answers.

Dear Mother and Father,

We are **(1)** .... of Captain Hook. We are on his big pirate **(2)** .... There are twenty **(3)** ....
pirates. It is very **(4)** .... at night. John is brave, but Michael is **(5)** .... We're very scared because we must **(6)** .... the plank at **(7)** ....
There's a big **(8)** .... waiting for us in the water!
John, Michael and I love you very **(9)** ....

Wendy

1 ☐ Wendy and the children are prisoners of ....
2 ☐ The pirate ship is ....
3 ☐ There are .... pirates.
4 ☐ It is very .... at night.
5 ☐ John is ....
6 ☐ Michael is ....
7 ☐ The children must walk the plank at ....
8 ☐ There is a big crocodile in the ....

A twenty   B water
C Captain Hook   D brave
E midnight   F crying
G big   H cold

BEFORE YOU READ

LISTENING

 Listen the first part of Chapter Seven and complete the text with the words you hear.

At the Darling (1) ...., Mr and Mrs Darling and Nana are desolate. They (2) .... think about Wendy, John and Michael. They look (3) .... the three empty beds and (4) .... come to their eyes. Mr and Mrs Darling never (5) .... or laugh anymore.

Mrs Darling (6) .... in the silent nursery and cries. She thinks (7) .... her children, their games and their happy voices. Nana tries to comfort her, but (8) .... can make Mrs Darling happy.

 What do you think happens next to Wendy, John and Michael? Choose from these answers or make up your own.

1 They become pirates with Peter Pan on the *Jolly Roger*.
2 They go back to Neverland and live happily with the Lost Boys.
3 They go home and the Lost Boys come with them.
4 They go home and Peter comes to live with them.

Fairies

12 Some people say that fairies:

- are little people with wings;
- live in the forest, in trees, in flowers and in gardens;
- are sometimes good and sometimes bad;
- they are magic;
- they can do good things and bad things.

Fairies are part of the British, Irish, northern European, Scandinavian and old Celtic cultures. There are many stories about fairies. Some are called fairy tales.

In *Peter Pan* Tinker Bell is Peter's fairy. The fairies of Neverland are Peter's friends. In *Pinocchio* there is a kind fairy called the Blue Fairy. She always helps Pinocchio. In *Cinderella* a fairy godmother is Cinderella's good friend.

There are fairies in one of William Shakespeare's plays, too. In *A Midsummer Night's Dream* Oberon and Titania are the king and queen of the fairies.

American and British children believe in the tooth fairy. When a child's tooth falls out he/she puts it under the pillow [1] and goes to sleep. During the night the tooth fairy takes the tooth and puts money or sweets under the pillow. The next morning the child finds the surprise and is very happy.

Dr W. Evans-Wentz is an expert on Celtic myths. He says, 'Fairies exist but they are invisible.'

What do you think? What do your classmates think?

1. **pillow** :

A Fairy (*c.* 1921) by Ida Rentoul Outhhwaite.

PROJECT ON THE WEB

Find out more about fairies on the Internet. Follow the instructions of page 32 to connect to the correct website. In small groups prepare a fact file on some different types of fairies. You can add some pictures if you like. Present your results to the class.

CHAPTER **SEVEN**

Home at Last!

t the Darling home, Mr and Mrs Darling and Nana are desolate. [1] They always think about Wendy, John and Michael. They look at the three empty beds and tears come to their eyes. Mr and Mrs Darling never smile or laugh anymore.

Mrs Darling sits in the silent nursery and cries. She thinks of her children, their games and their happy voices. Nana tries to comfort her, but nothing can make Mrs Darling happy.

One night after several months, something incredible happens. Wendy, John and Michael fly into the nursery! Mrs Darling is sitting near the fireplace.

'Mother, mother, we're home!' says Wendy.

Mrs Darling turns around and sees her three dear children.

'Is this true or is it a dream? I can't believe it!' she says.

'Oh, Mother, we're home at last,' the children say.

1. **desolate** : very sad.

Wendy, John and Michael embrace their mother and kiss her.

'How wonderful to see you, my dear children! How wonderful to hear your sweet voices. Oh, let me look at you!' She calls Mr Darling. Mr Darling is very happy and surprised.

There is great joy in the Darling nursery tonight.

'Mother,' says Wendy, 'Peter Pan and the Lost Boys are here too. They are waiting outside.'

The six Lost Boys slowly enter the nursery. They look at Mrs Darling and smile at her.

'Mother, these are the Lost Boys. They haven't got a mother. Can they stay with us?' says Wendy.

'What dear little boys!' says Mrs Darling. 'Of course they can stay with us. And where is Peter Pan?'

Peter enters the nursery and says, 'I'm here, but I don't want to stay here. I don't want to go to school and I don't want to grow up! I want to be a young boy forever. I must return to Neverland. I'm happy with the Indians and the fairies.'

Wendy is surprised and says, 'But Peter, when will I see you again?'

Mrs Darling says, 'I have an idea. Wendy, you can visit Peter in Neverland every spring! You can stay there for a week.'

'Can I really go to Neverland every spring, Mother?' asks Wendy.

Peter looks at Mrs Darling and asks, 'Is that a promise?'

'Of course it is,' says Mrs Darling.

'Then I want spring to come quickly,' says Peter.

'Yes, very quickly,' says Wendy.

'Come on, Tink! Let's fly home and wait for spring,' says Peter.

Peter Pan and Tinker Bell fly out of the nursery window into the night sky. Their destination? [1] Neverland!

---

1. **destination** : the place where they are going.

UNDERSTANDING THE TEXT

1 **Put the words in the correct order to make questions about Chapter Seven. Then answer them.**

1 Mr and Mrs Darling/are/why/sad?
2 Darling children/arrive/do/when/the?
3 is/the/with/who/three children?
4 to live/where/Peter/does/want?
5 Wendy/and/Peter/when/go/can/visit?
6 fly/Peter/with/does/home/who?

2 **TELLING THE STORY**
Number the pictures in the correct order to make a summary of the story. Then describe what is happening in each picture.

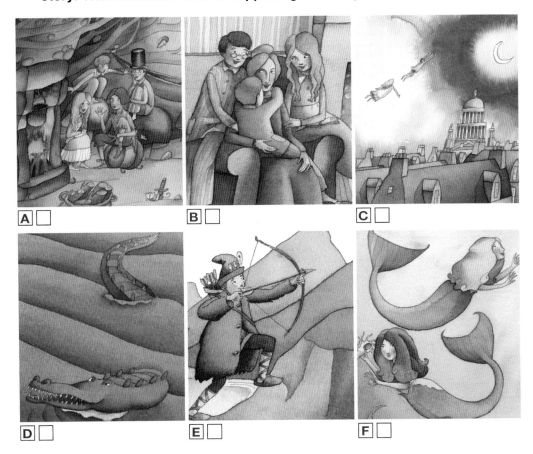

A ☐ B ☐ C ☐

D ☐ E ☐ F ☐

A MAZE

Help the Lost Boys go through the maze. Circle the objects that start with the letter 't' and try to reach the treasure chest. Do you know the names of the other objects in the maze?

Peter Pan in Films and at the Theatre

Peter Pan on film:

- *The Adventures of Peter Pan* (1953) by Walt Disney.
- *Hook* (1991) by Steven Spielberg.
- *Finding Neverland* (2004) by Marc Forster.

Johnny Depp, Kate Winslet and other actors
in a scene from *Finding Neverland*.

PROJECT ON THE WEB

Find out more about the films. Follow the instructions on page 32 to find the correct website.

▶ Read the information about the film and the characters.

▶ After you read the information, write some information about each of the characters in *Peter Pan*.

▶ Play the Captain Hook game. Have fun and practise your English!

Peter Pan at the theatre:

1 **Which phrase (A-E) says this (1-5)? For sentences 1-5 write the correct letter in the box.**

1 ☐ We have evening performances.
2 ☐ We're closed on 24, 25, 31 December and 1 January.
3 ☐ Children have special rates.
4 ☐ We have afternoon performances.
5 ☐ We have special family tickets.

A Matinees at 2.30 p.m.
B Special children's rates.
C Evenings at 7 p.m.
D Family packages available.
E No performances on 24, 25, 31 December and 1 January.

EXIT TEST

1 **Are these sentences true (T) or false (F)? Correct the false ones.**

		T	F
1	Michael likes having baths.	☐	☐
2	Tinker Bell finds Peter's shadow in the sewing box.	☐	☐
3	There are six Lost Boys.	☐	☐
4	The pirates and the Indians are friends.	☐	☐
5	Tiger Lily is a prisoner of the pirates.	☐	☐
6	The Underground Home has a lot of small rooms and a fireplace.	☐	☐
7	The *Jolly Roger* is in the bay near Kidd's Creek.	☐	☐
8	John, Michael and the Lost Boys walk the plank.	☐	☐
9	Wendy can visit Peter in Neverland every spring for a week.	☐	☐

2 **CHARACTERS**
Do you recognise them? Read the description and write the name of the character.

1 He is a wicked pirate captain. ....

2 She is a beautiful Indian girl. ....

3 She is a lovely fairy. ....

4 He is a pirate. ....

5 She is the children's nanny. ....

6 He wants to be a boy forever. ....

7 She tells bedtime stories to the Lost Boys. ....

8 He is the second child of the Darling family. ....

9 He is the third child of the Darling family. ....

Who is your favourite character? ....

Why? ....

Use the words in the box to complete the summary.

home	fights	saves	happy	stories	plank	night
pirates	Neverland	swim	capture	London	daughter
return	meet	mother	crocodile	live	Underground

Wendy, John and Michael **(1)** .... in London. One **(2)** ....
Peter Pan comes to their nursery. They fly with him to **(3)** ....
They **(4)** .... the Lost Boys. They live in the **(5)** .... Home.
Wendy is a perfect **(6)** .... She tells them **(7)** .... and they
are very **(8)** ....

Wendy, John and Michael **(9)** .... in Mermaids' Lagoon. Tiger Lily
is the **(10)** .... of the Indian chief. Peter **(11)** .... Tiger Lily
from the pirates.

Captain Hook and his **(12)** .... live on the *Jolly Roger*.
One evening Wendy, John and Michael decide to return **(13)** ....
But the pirates **(14)** .... them.

Captain Hook says, 'You must walk the **(15)** ....!'
Peter Pan arrives and **(16)** .... with Captain Hook. Hook falls
into the mouth of the **(17)** .... Peter takes the children to
(18) .... He and Tinker Bell **(19)** .... to Neverland.